THE GARDEN ROOM

Priscilla Boniface

GENERAL EDITOR
Peter Fowler

ROYAL COMMISSION ON HISTORICAL MONUMENTS ENGLAND
LONDON HER MAJESTY'S STATIONERY OFFICE

© Crown Copyright 1982
First published 1982

ISBN 0 11 701127 4

HER MAJESTY'S STATIONERY OFFICE

Government Bookshops

49 High Holborn, London, WC1V 6HB
13a Castle Street, Edinburgh EH2 3AR
41 The Hayes, Cardiff CF1 1JW
Brazennose Street, Manchester M60 8AS
Southey House, Wine Street, Bristol BS1 2BQ
258 Broad Street, Birmingham B1 2HE
80 Chichester Street, Belfast BT1 4JY

*Government publications are also
available through booksellers*

SELECT BIBLIOGRAPHY

Jill Franklin, *The Gentleman's Country House and Its Plan. 1835–1914*. Routledge, Kegan Paul 1980
John Harris (ed) and Hugh Johnson (intro). *The Garden*. Mitchell Beazley and New Perspectives Publishing Ltd. 1979
Shirley Hibberd, *The Amateur's Greenhouse and Conservatory*. London 1897
Robert Kerr, *The Gentleman's House*. London 1871
John Claudius Loudon, *Remarks on the Construction of Hot Houses*. London 1817
Hermann Muthesius, *The English House*. Granada. Crosby, Lockwood Staples. 1979
Walter Nicol, *The Forcing Fruit and Kitchen Gardener*. Edinburgh 1802

front cover Winter Garden, Horncliffe House, Northumberland. NMR 1981. *inside front cover* NMR, 1981. Copyright Boulton and Paul Limited. *inside back cover* NMR, 1981. Copyright Boulton and Paul Limited. *Frontispiece* The Conservatory, Alton Towers, Staffordshire. F. J. Palmer, 1951. Copyright NMR.

[Printed in England for Her Majesty's Stationery Office by
W. S. Cowell Ltd Butter Market Ipswich Dd 717080 C60]

EDITOR'S FOREWORD

All the photographs in this book are held in the National Monuments Record (NMR), a national archive which is part of the Royal Commission on Historical Monuments (England). The NMR originated in 1941 as the National Buildings Record which, at a time when so much was being destroyed, took upon itself the task of photographing as many historic buildings as possible before it was too late. The Record continued its work after the War and was transferred to the Royal Commission in 1963. As the NMR it now covers both architectural and archaeological subjects and contains well over a million photographs, together with maps, plans and other documents, relating to England's man-made heritage. The NMR is a public archive, open from 10.00 – 17.30 on weekdays; prints can be supplied to order on payment of the appropriate fee.

This book is the third of a series intended to illustrate the wealth of photographic material publicly available in the NMR. Many of the photographs are valuable in their own right, either because of their age or because of the photographer who took them or because they are the only records we now possess of buildings, and even whole environments, which have disappeared. Unlike other Commission publications, these are primarily picture-books, drawing entirely on what happens to be in the NMR. There is no attempt to treat each subject comprehensively nor to accompany it with a deeply researched text, but the introduction and captions are intended to give meaning to the photographs by indicating the context within which they can be viewed. It would be pleasing if they suggested lines of enquiry to others to follow up.

The early titles in the series show where the strengths of the archive lie. Equally, of course, the collection is weak in some respects and I hope that many of those who buy this volume may be reminded of old, and perhaps disregarded, photographs of buildings in their possession. We would be glad to be told of the whereabouts of such photographs as potential contributions to a national record of our architectural heritage.

This particular collection of photographs is, like other books in the series, ostensibly about buildings. The types illustrated here are, however, of a rather special type linked together by the idea of a 'garden room', that is not so much a room in the garden (though there are examples of such here) as a room which brings the garden into the house. The furniture of such rooms was as that of a garden, so some trouble has been taken here to identify and discuss the flora in them rather as would be done if the 'buildings' were in fact a garden.

Royal Commission on Historical
Monuments (England),
Fortress House,
23 Savile Row,
London W1X 1AB

Peter Fowler,
Secretary,
Royal Commission on Historical
Monuments (England);
General Editor,
NMR Photographic Archives

ACKNOWLEDGEMENTS

The Commission is grateful for permission to reproduce photographs in the National Monuments Record of which the copyright is held by:

Richard Booth
Boulton and Paul Ltd.
Michael Brand
British Museum
Captain G. Brodrick
B.W.S. Publishing (*Architect and Building News*)
City of Nottingham Arts Department
Clinton Devon Estates
C. L. S. Cornwall-Legh
Howarth-Loomes Collection
Pilgrim Trust
C. Sebag Montefiore
Christopher Wood Gallery

The author would like to thank many individuals and institutions for help, in particular: Andrew Adam and Julia Lacey (Boulton and Paul Ltd.), Col R. A. Alec Smith, R. T. Brocklebank (Birkenhead Central Library), Lady Bromley Davenport, D. Butler (County Record Office, Durham), C. L. S. Cornwall-Legh, K. A. Doughty (Southwark Borough Librarian and Curator), Dr B. Elliot (Lindley Library, Royal Horticultural Society), Mrs G. Hibbert (ICI). M. Holmes (Local History Library, Swiss Cottage), J. T. Hopkins (Cheshire Record Office), D. J. C. Howard (Burhill Estates Co. Ltd.), Mrs A. McCormack (Surrey Record Office, Kingston upon Thames), B. R. Playle (Curator, Natural History Museum, Wollaton Hall), Dr C. A. Ralegh Radford, N. J. Reading (Architects Department, Historic Buildings Division, Greater London Council), P. M. Reid, W. Serjeant (County Archivist, Bury St Edmunds), Mrs C. Short (Cultural Activities Centre, Sheffield), Miss Lucy Strickland, R. J. Waller (Clinton Devon Estates).

It is a pleasure to give particular thanks to Mr. R. H. Butcher, Internal Information Officer of Colman Foods, who has been of tremendous help.

Colleagues on the Commission staff also deserve thanks, especially Miss J. Carden, Mr. R. Flanders, Mrs. D. Kendall and the staff in the Order Section, and Mr. R. Parsons and staff in the Photographic Section.

Many members of staff at HMSO have enthusiastically given assistance. Mr G. Warren has designed the book and steered it through its various stages with skill, patience and good humour. Mr David Joyce, Editor at HMSO, has made many improvements to the text and his advice and interest have been of enormous benefit to the series.

Finally acknowledgement should be given to Dr Peter Fowler, the General Editor, for the encouragement and guidance which he has given to all the authors in the series.

THE GARDEN ROOM

INTRODUCTION

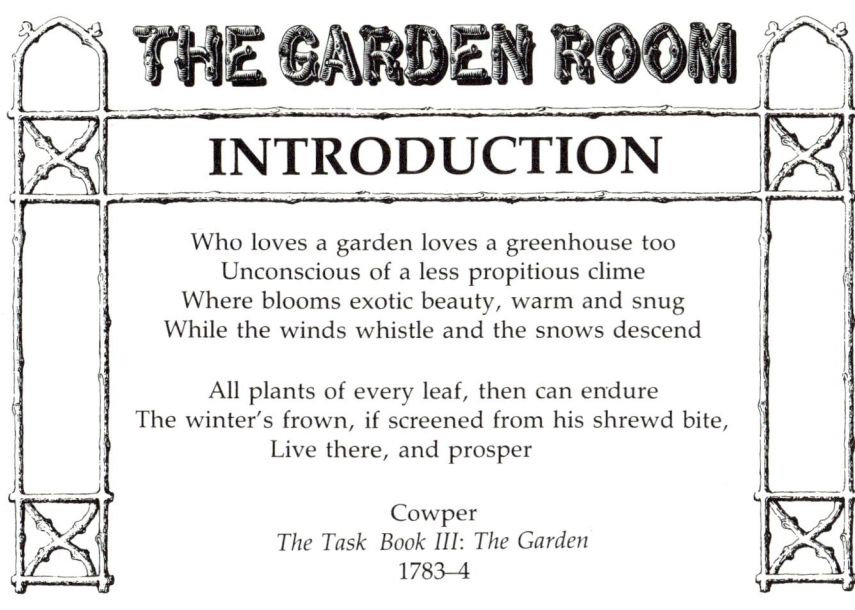

> Who loves a garden loves a greenhouse too
> Unconscious of a less propitious clime
> Where blooms exotic beauty, warm and snug
> While the winds whistle and the snows descend
>
> All plants of every leaf, then can endure
> The winter's frown, if screened from his shrewd bite,
> Live there, and prosper
>
> Cowper
> *The Task Book III: The Garden*
> 1783–4

A garden room may be defined as any accommodation for which the garden and more particularly plants are the *raison d'etre*. The term encompasses rooms as diverse as summer houses and orangeries, greenhouses and loggias, winter gardens and tree-houses. Pre-eminent among them is that largely Victorian phenomenon, the conservatory, which incidentally mirrored social change. The idea, however, of a garden room might be said to date from three centuries earlier in 1590 when the orange tree was first brought to Britain; indeed experimenting with hot-houses had started on the continent as early as the 12th century.

Precursors of the conservatory
The first houses built for plants in Britain were orangeries. In 1629 John Parkinson writing in *Paradisi in Sole Paradisus Terrestris* recommended placing orange trees 'in a house or close gallery in the winter time'. Orangeries were heated by an open fire or a coal- or wood-burning stove that had a flue in the back wall or floor of the building. One of the earliest surviving orangeries was erected in 1704 at Kensington Palace. The appearance of innovations at Royal Palaces, where there were sufficient foreign visitors to stimulate the introduction and implementation of new ideas, was not unusual.

By 1660 it had been suggested that vines or apricot trees should be grown near a wall heated by a kitchen, or other, fire. The fuel was coal, but a cheaper way of providing heat was by piling dung against the wall. Phillip Miller, Gardener at the Chelsea Physic Garden (which was started in the 1670s), wrote in his *Gardener's Dictionary* (1740) that the use of hot walls for ripening fruits 'is pretty much practised in England'. Loudon, claiming in *Remarks on the Construction of Hot Houses* (1817) that the Duke of Rutland at Belvoir was the first to adopt hot walls about or before 1710, said that the purpose of planting beside hot walls was to mature young wood and fruit in autumn and to protect blossom from frosts in spring. The hot wall method was suitable for the propagation of cherries, plums, grapes, apricots, peaches, nectarines, pomegranates, oranges, lemons and melons. In front of any of these plants, strawberries, dwarf peas and beans could be forced in pots, often protected by glass covers or oiled paper sheets. Loudon observed that hot walls had become general in the north of England and Scotland.

The word 'stove' was used either to describe that object as we understand it or the room in which it was housed. Watts, Gardener to the Society of Apothecaries, is credited with being, in 1684, the first person to convey heat under the floor *of a greenhouse* (though the method had been well-known in Britain in Roman times). John Evelyn visited the Society's garden in Chelsea on 7 August 1685, and recorded: 'what was very ingenious was the subterranean heat, conveyed by a stove under the conservatory, all vaulted with brick, so he has the doors and windows open in the hardest frosts, secluding only the snow'.

Lord Petre had installed three huge stoves (i.e. hothouses) at Thorndon Hall, Essex, by 1736, where fruit such as pineapple, guava, ginger and lime, and also the first camellia in England, were successfully grown. Bark-fuelled stoves were used for growing date, coffee, cocoa, cashew, pimento, vanilla and banana. Dry heat stoves, with plants on scaffolding, proved useful for rearing succulent plants like aloe. The pineapple was introduced to Britain in 1719 and came to be the yardstick by which the talents of a gardener like Barnes of Bicton were judged. The fashion for growing pineapples died when the country was flooded with cheap colonial imports in the 1860s.

Hot beds were certainly in use by 1715; they were mentioned by Laurence in the last edition of his *Kalendar*. Dung, leaves or bark that heated up naturally were used to fill the beds which, like hot walls, were protected by glass or oiled paper covers. Melon and cucumber grew best in dung. Philip Miller had successfully grown Barbadan coconut and other nuts in pots of tanner's bark by 1728.

Few further improvements were made in the artificial forcing or protection of plants until the beginning of the 19th century. Until then there was no widespread demand for early and highly flavoured fruits, although individuals like Richard Payne Knight at Downton, Herefordshire, had continued experimenting with techniques for forcing plants.

Conservatories
In the early decades of the 19th century a number of circumstances led to the development of the 'Victorian Conservatory'. The new fashion for a 'picturesque' style of architecture allowed an asymetrical house plan which could easily incorporate a conservatory. A leader of the 'picturesque' movement, John Nash, attached one of the first conservatories (in fact named a 'viranda') to a house at Luscombe in Devon in 1799.

Definitions vary as to what constituted a conservatory and a winter garden. Shirley Hibberd speaking of stoves, glasshouses and conservatories said that they were all plant houses, and depended for their distinctions quite as much on the furnishing and the management as upon structure and fittings. Stevenson says a conservatory was 'not for raising plants, but for exposing them when in bloom'. Robert Kerr thought it all a matter of size. He said the term 'winter garden' was applied to a glasshouse 'say 50 feet square or upwards'. He went on to say 'the purpose in every such case is to accommodate, for gardening effect rather than mere conservation, a collection of rare plants to be kept in condition during winter by artificial heat, interspersed with sculptures, rockwork, shellwork, one or more fountains and so on, and the pillars shrouded in masses of creepers and pendant runners'.

What does not seem to be in doubt is that a conservatory or winter garden was supposed to be entertainment. Among the flowers were walkways for promenades, perhaps a distant descendant from the long glazed galleries of Tudor houses, which allowed protected walks to be taken while looking out to the grounds outside. Usually seats were provided for reading, lounging or taking light refreshment. Conservatories were often lit by gas, chinese lanterns or, latterly, by electricity to allow evening walks. A conservatory was often used for a party as at Wollaton Hall and Capesthorne Hall. Knight's *Cyclopaedia of London* (*c.* 1851) described the new Winter Garden in Regent's Park: 'From the keen, frosty air outside, and the flowerless aspect of universal nature, one steps into an atmosphere balmy and delicious and not in the slightest degree oppressive. The most exquisite odours are wafted to and fro with every movement of the glass doors. Birds singing in the branches. . . .make you again and again pause to ask, is this winter? Is this England?'

The mention of birds in a winter garden sounds extraordinary. Nevertheless there were alleged to be birds in the Great Stove at Chatsworth and white doves appear in a painting of the conservatory at Wood End, Scarborough, home of the Sitwells.

Up to 1800, few conservatories had been built of materials other than masonry or wood. Subsequently, improved iron-casting techniques resulted in the construction of buildings which previously could not have been attempted. So far as is known, the first complete glass and iron façade was built at the Camellia House, Wollaton Hall, Nottinghamshire, in 1823.

Paxton, head gardener at Chatsworth, and John Claudius Loudon both experimented with various shapes of glasshouse in attempts to gain the maximum heat from the sun. Loudon, who had avowed 'to make English hot-houses look beautiful in their own right instead of being merely lean-to glazed sheds', read a paper before the Horticultural Society in 1815 stating that the most perfect form of hothouse was a glazed semi-globe. He and Paxton both came to favour the ridge-and-furrow construction for roof glazing which was thought to catch the sun's rays at a right angle and, Loudon said, accomplished 'two daily meridians'.

Glazing at the beginning of the 19th century was still very expensive. Previously, roofs of orangeries had usually been of lead or slate. Walter Nicol, writing in *The Forcing Fruit and Kitchen Gardener* (1802), advised the use of small panes of glass; 'the price of the superficial foot of glass varying according to the size of the squares, it is of importance not to make these too large'. He stated that a 12 inch square would cost ten pence whereas two 8 inch squares could be had for six pence halfpenny because they could be cut from the waste of other cuttings.

In 1832 Chance Bros introduced a method of blowing cylinder glass that had originated in Germany. It now became feasible to manufacture bigger sheets than hitherto. With glass tax removed in 1845 and window tax repealed in 1851, a rush started to own a conservatory.

A key to the development of the conservatory was a marked improvement in the methods of heating. In 1802 Walter Nicol suggested using two stoves (ovens) rather than one to spread heat more evenly. It was then discovered that steam heat passing through cast-iron water pipes allowed an even distribution of heat and a more constant temperature without the attendant dust, smoke and smell of hydrogen gas leaking from flues. Shirley Hibberd in *The Amateur's Greenhouse and Conservatory* (1897) prescribed heating a glasshouse by hot water or hot air but thought hot water better because there was more attendant moisture. If a conservatory was attached to a small house, hot water could be piped from the kitchen stove.

The siting of a stove, greenhouse or conservatory was of vital importance in relation to the heating. Obviously the more warmth obtained from the sun, the less was needed from artificial sources. There were basically three types of greenhouse; the lean-to, the three-quarter span and the span. Advice on their location was along the following lines.

The lean-to should be against a south-facing wall so that the sun would be on the east end in the morning, all over the house at midday and at the west end in the afternoon. A three-quarter span house was best placed on the west side of a wall. In this way the front facing west would get the greatest amount of sun and the house would therefore require less heating at night. A span house should be on a north-south axis. If ferns were to be cultivated then the length of the house should extend east-west and the ferns should be placed on the north side of the house where they would be in the most shade. To conserve fuel at night canvas or mats would often be placed over the roof. Most conservatories were equipped with blinds, which served a similar function and at the same time discouraged condensation.

A greenhouse or conservatory needed plenty of ventilation and it was recommended by Robert Kerr that 'sashes ought almost all to be open'. The temperature of a house depended very much on what plants it contained. A stove housing orchids or pitcher plants was heated to 80 or 90 degrees Fahrenheit. Pineapples required a heat of 65 degrees until 1 March, 70 degrees thereafter. A botanical thermometer was used in their cultivation which had to be raised to the Anana mark. On the other hand, a temperate climate of 40 to 50 degrees was sufficient for flowers such as camellias, acacias and pelargoniums. Ferns and alpines, needing no heat at all, merely required the protection given by a greenhouse. Paxton commented, 'It not infrequently happens that when a plant is first introduced into this country it is cultivated at a much higher temperature than is really necessary for it'. Robert Kerr advised 'forcing of any kind being an outrage done nature, the more we avoid it the greater will be our success. She should be kindly assisted, not spurred nor thwarted when it can be avoided.'

There was an acceleration in the vogue for collecting exotic plants during the 19th century. The introduction of 'the Wardian Case' (like a miniature greenhouse) by Nathaniel Bagshaw Ward in 1833 made more successful the long-distance transportation of many plants which would until then have died.

The owners of newly discovered and imported plants naturally wished to show them off in the most flattering and glamorous surroundings. They would therefore strive to boast a number of greenhouses (usually connecting to allow the use of one heating system) in which to cultivate plants, and a conservatory or winter garden in which to place the plants for display. A large estate would probably have houses for muscat grapes, melons and cucumbers, mushrooms, bananas and pineapples, figs and nectarines, cherries and plums, ferns, azaleas, hyacinths, and orchids. A small estate would at least have houses for grapes (one for early, one for late crops), flowers and palms, and two peach houses.

The flowers in a conservatory were changed with the seasons. In a heated conservatory there might be at various times of the year: climbers such as bougainvillaea and plumbago; foliage like 'India rubber' plants, maidenhair fern and tradescantia; flowers such as acacia, azaleas, camellias, clematis, begonias, salvias, geraniums, chrysanthemums, arum lilies, hyacinths, tulips and narcissus; and, in addition, wall and basket plants. The temperature ranged between 45 and 55 degrees Fahrenheit at night and between 55 and 60 degrees during the day. An unheated conservatory might contain: foliage such as harts-tongue fern and yucca; flowers like auricula, primula, passion flowers, japonica, Christmas rose, hydrangea, agapanthus, roses and fuchsia with camellia and plumbago on the walls and campanula in hanging baskets. Cold house plants include palms, bamboos, yuccas; and herbaceous and alpine plants that flower during winter and spring. Shirley Hibberd recommended using annuals in pots if a cheap display was required. Of plants used for background greenery, anything with variegated leaves found particular favour with the Victorians. Pests were often a nuisance. They were dealt with by fumigation.

The construction of a conservatory or winter garden, the bringing of the garden to the house, was not without its problems. Although Loudon thought conservatories 'useful in a medical point of view', several writers give dire warnings of their dangers e.g. 'to be too directly attached to a Dwelling Room is inadvisable. . . the warm moist air, impregnated with vegetable matter and deteriorated by the organic action of the plants, is both unfit to breathe and destructive of the fabrics of furniture and decoration' (Robert Kerr in The *Gentleman's House* 1875); 'to avoid damp and the smell of earth in the house, it is well to interpose a porch or corridor' (J. J. Stevenson in *House Architecture* 1880). A conservatory usually led from a drawing-room, boudoir, or morning room but via a saloon, vestibule, gallery or corridor which acted as buffer. Stevenson bravely allowed that one of the drawing-room windows might look into a conservatory. At Abbey Manor, Evesham, a false mirror in the dining-room gave a view through to the conservatory but most house owners, like Sir Morton Peto at Somerleyton, seemed content to follow Stevenson's advice and have a passage between house and conservatory. Indeed, about one third of all conservatories built had no access from the house at all; entrances were from the garden only.

Social trends
By tracing the history of the conservatory one may also note the shift of influence during the 19th century from the hereditary land-owners to the *nouveaux-riches* industrialists and professional men.

Cheap corn imports in the 1870s may perhaps be identified as one of the prime causes of the reduced circumstances of many members of the landowning aristocracy in the later Victorian period. Certainly the events of that decade hastened the rise of 'the self-made man' to positions of power based on property and income. A writer at the time put it romantically: 'the Able man of the Middle Ages was he who gathered round him a devoted band of followers to conquer his fellows by force of arms and wrest from them an abundance of spoil; in our time the Able man is the Captain of Industry, gifted with the faculty of organizing and governing human endeavour to win conquests in the peaceful fields of commerce'. One incidental result of this development, allied with improved techniques of mass production, was that conservatories ceased to be the prerogative of the 'old' landowning families; conservatories in fact went 'down market'. The glass palaces of the mid-Victorian period were no longer built, but many of the glass houses that were erected were if anything more luxurious and ostentatious. This suited the psyche of the rich men who wanted to proclaim their new-found wealth at both their town and week-end houses.

The parvenues continued the tradition of *noblesse oblige*. They were often intensely philanthropic and a surprisingly large number became Members of Parliament. At the turn of the century, South African gold and diamond merchants and men who had made fortunes in the New World became a part of the English social scene. The Prince of Wales preferred the company of the *arrivistes*, a regal patronage which hastened their acceptance by the Establishment.

The Victorian Summer House
The summer house, a completely different type of garden, or 'inside out', room was also popular during the 19th century, providing a pleasant area for sitting, taking tea and looking out to the garden. A number of architectural styles were adopted for the Victorian summer house ranging from the West Indian to the Japanese, the Swiss to the Zulu, the choice often reflecting the origins of the plants that surrounded them. The predominate style, however, was 'rustic' to emphasize the natural, less formal use which was made of the garden in Victorian and Edwardian England.

The decline of the conservatory
By 1900, although a prefabricated conservatory was being added to nearly every small middle-class house, its apogee was already over in fashionable circles. The 'Arts and Crafts' movement disapproved of so artificial a barrier between house and garden. The loggia, formerly popular in the early 17th century, was to prove an acceptable substitute. Already in the 1880s a loggia had been built at Clouds, Wiltshire, and patrons and collectors who moved in 'artistic' circles soon followed suit. The only architect with 'Arts and Crafts' ideals who is known to have designed a conservatory was Philip Webb at Standen; and there, it is thought, it was at his client's request. Edwin Lutyens and the neo-Classicists favoured a loggia, attached to a house or in the grounds, which could still afford some personal protection. Gradually, however, the Edwardian ideal of a lily-white pallor gave way to the health and beauty cult. This positively encouraged sun and wind on the skin and that most un-Victorian concept, a 'tan', the proud owners of which were labelled in France *'les petites chocolatières'*.

Nevertheless, many conservatories continued in use until the First World War. Then, however, in a period of such great social and economic changes as to prove a watershed in so much of the English way of life, the losses in the trenches of so many of the men who were the work force to maintain them and a sharp increase in the cost of fuel resulted in the demolition of large numbers of glass 'garden houses' of different types. Others were just left to decay. The magic world of the winter garden described by Loudon was gone: 'winding walks, fountains, and even plots of grasses and ponds of water, so that the only difference between them and real gardens is, that glass intervenes between the summit of their trees and the sky; and nothing can be more delightful when there is frost and snow upon the ground outside, than to enjoy the general warmth and verdant beauty within'.

1 THE ORANGERY, BARTON SEAGRAVE HALL, BARTON SEAGRAVE, NORTHAMPTONSHIRE.
Early hot houses in England were used for the protection during bad weather of orange trees: they were moved outside during the summer. 'Glazed architecture' orangeries provided sufficient light for the purpose since the aim was to protect rather than to encourage growth.

The late 18th century Orangery at Barton Seagrave represents a stage in the gradual development of buildings that would admit more light beneficial to the plants. Unusually for the time, cast iron was used for the elongated columns along the façade. G. B. Mason, 1946.

2 THE ORANGERY, SEZINCOTE, GLOUCESTERSHIRE.
The house was built c. 1805, a time when the exploits of Clive and Warren Hastings were still in the forefront of public knowledge. In addition, Thomas and William Daniell had published drawings of Indian architecture in 1795 which drew attention to a 'Hindu' style.

The estate was bought by a nabob, Colonel John Cockerell, Quarter Master General under Lord Cornwallis. John Cockerell's brother Charles had also lived in India. Another brother, the architect S. P. Cockerell, was Surveyor to East India House. Sezincote was the joint creation of Charles and Samuel Pepys Cockerell, Thomas Daniell and Humphrey Repton.

The drawing by John Martin, who had been commissioned to make an illustrated record of Sezincote, shows the curved orangery at the left of the house. The orangery is of cast iron, stained glass and Coade stone. The Wellington Memorial in the grounds of Sezincote is in fact the heating chimney of the orangery. Copyright British Museum.

3 THE ORANGERY, PANSHANGER, HERTINGFORDBURY, HERTFORDSHIRE.
Panshanger was altered several times during the 19th century: in 1806 by William Atkinson, in 1819, and in 1855 following a fire. Repton landscaped the grounds. The orangery is not unlike that at Castle Ashby, Northamptonshire.

Panshanger, under the ownership of the 5th Earl Cowper and latterly the Pagets, was famous for its house parties. Queen Victoria, Lord Melbourne and Mr Balfour were among visitors to the house. A group of intellectuals named the Souls, which included George Curzon, Herbert Asquith and George Wyndham, met here for charades, discussion etc. The house was described as 'full to the brim of vice, agreeableness, foreigners and roués'. Mr and Mrs Almeric Paget purchased the house in 1917 and continued the tradition of house parties. John Buchan met his wife here. The last occupant was Lady Besborough, who died in 1952. The house was demolished in 1953–4. Newton, c. 1910

4 (*opposite, below*) THE ORANGERY, WREST PARK, SILSOE, BEDFORDSHIRE.
The Orangery, a most remarkable building for its date, 1836, is thought to have been designed by the owner of Wrest, the 2nd Earl Grey, with the assistance of his Clerk of Works, James Cléphane. Cléphane, a Frenchman who had worked for Lord Barrington at Becket Park, Berkshire, was paid only a weekly salary by Earl Grey, an arrangement which would appear to underline his minor status in the construction.

Earl Grey, a passionate Francophile, was influenced in his design by a book entitled *Architecture Française* written by Jacques Françoise Blondel in 1752, which was a study of early Louis XV architecture at Paris and Versailles.

The shrubs and plants were supplied by King Louis-Phillippe. Newton, *c.* 1910.

5 THE ORANGERY, BYRAM PARK, BYRAM CUM SUTTON, NORTH YORKSHIRE.
The Orangery dates from the late 18th century and is thought to have been designed by John Carr of York. A mid 18th-century door from Methley Hall was re-used in the building. The glazed extensions and roof must be later additions.

The hall, which was the home of the Pennington-Ramsden family, was mostly demolished in the 1920s. Day, *c.*1910

6 (*overleaf*) THE ORANGERY, CASTLE ASHBY, NORTHAMPTONSHIRE.
This building, designed by Matthew Digby Wyatt in 1861–5, indicates that the fashion for 'glazed architecture' did not die out completely after the repeal of the glass tax in 1845. The use of glass with as little visible support as possible obviously did not have universal appeal. Bedford Lemere, 1892.

7 GREENHOUSE, EAST CLIFF LODGE, RAMSGATE, KENT.
East Cliff Lodge, the home of the Sebag-Montefiores, city merchants, is thought to have been designed by Decimus Burton in 1831–2.

The design of the greenhouse must have been influenced by the experiments of Loudon and Richard Turner, which sought to gain optimum warmth from the sun's rays by varied yet precise positioning of the angle of the glass panes. Photographer unknown, c. 1920. Copyright C. Sebag-Montefiore.

8 GREENHOUSE, FARLEY HALL, FARLEY, STAFFORDSHIRE.
This greenhouse is almost a perfect semi-globe as recommended by Loudon in his remarks to the Horticultural Society in 1815. It stands on a stone wall that joins the main house to a Gothic billiard room built in 1866. NMR, 1979.

9 ORANGERY, MAPLEDURHAM, OXFORDSHIRE.
Mapledurham was owned by the Darell-Blount family. According to tradition, whenever a member of the family died a tree in the avenue in the grounds fell.

This orangery was a glasshouse with minimal masonry supports. It was strategically placed to close the view from the east front of the house. Newton, c. 1900.

10 (*opposite*) **ORANGERY, HOWSHAM HALL, HOWSHAM, NORTH YORKSHIRE.**
Howsham Hall was the seat of the Fairfax-Cholmley family. It ceased to be a private house in 1948 and soon afterwards became a boys' preparatory school.

The outstanding feature of this brick orangery is the delicate glass roof. The date of the building is not known but the tiny panes of glass would indicate that it was erected before 1832 when blown cylinder glass was introduced in England, and before 1845, when the Glass Tax was repealed. Herbert Felton, 1953.

11 GREENHOUSE, BROMBOROUGH HALL, BEBINGTON, MERSEYSIDE.
Professor Pevsner described the village of Bromborough as a 'mid Victorian Eden of large houses'. Bromborough Hall was the home of Sir William Forwood, who was influential in the building of Liverpool Cathedral. He described Bromborough Hall in his book *Some Recollections of a Busy Life* (1910).

'It is a very old house built in 1617, but enlarged several times since. It is partly Georgian and partly Old English Homestead. The entire south front is wreathed with wisteria, jasmine and clematis, making it harmonise with the Old Dutch garden that stretches before.

The gardens have an extent of about thirteen acres, and contain probably the most extensive lawns and largest trees in Wirral. The outlook across the Mersey is extensive and very lovely.'

The house was demolished in 1932 and shortly afterwards a road was cut through the grounds, which then became overgrown and were known as 'the Village Wood'.

The greenhouse is representative of those manufactured at the turn of the 19th century. Day, *c.* 1910.

12 (*opposite, top*) **CAMELLIA HOUSE, WOLLATON HALL, NOTTINGHAM, NOTTINGHAMSHIRE.**
The Camellia House at Wollaton, built in 1823, is thought to have been the earliest example of a complete cast iron and glass façade in England.

The *Nottingham Review* and *Nottingham Journal* of February 1824 describe a party at Wollaton when 'the brightness of the sky gave an opportunity of using the beautiful conservatory which had recently been erected by his Lordship, and is filled with the choicest of plants'.

When first introduced into this country in the early 19th century, the camellia was often put in too hot a house. Mawe and Abercrombie in *Universal Gardener and Botanist* (1797) said of the camellia 'though it is retained here as a greenhouse plant, there is no doubt, if trained to a south wall and protected from severe frost, there will be little danger of its being destroyed in winter where it will exhibit a profusion of its charming flowers . . . in the Spring'.

Camellias still occupy the Camellia House, which is at present (1981) being renovated. Copyright City of Nottingham Natural History Museum, 1970s.

13 (*centre*) **THE PALM HOUSE, BICTON HOUSE, BICTON, DEVONSHIRE.**
In the early 19th century, Lord John Rolle, a keen horticulturalist, lived at Bicton, bringing to it many plants he had collected on the continent.

James Barnes, a gardener at Bicton from 1839 to 1869, was a regular contributor to gardening magazines. He was famous for his pineapples. Barnes cured rust on grape vines and evolved a new strawberry strain and a type of garden broom.

The cast-iron framed Palm House is thought to have been erected *c* 1820–5. James Barnes raised the roof! The house now contains mimosa, passion flower, heavenly bamboo and the tea tree from New Zealand. Copyright Clinton Devon Estates, 1970s.

14 (*bottom*) **CONSERVATORY, ENVILLE HALL, ENVILLE, STAFFORDSHIRE.**
A glass palace erected by Grey, Ormson and Brown for the Earl of Stamford and Warrington in 1854.

The glasshouse was heated by boilers below ground level. A smoke tunnel ran 100 yards underground to end disguised as the corner tower on the 'Gothic' Museum.

A demolition team stationed at Enville during the Second World War blew up the glasshouse for practice. Photographer unknown, *c*. 1910

THE GREAT STOVE, CHATSWORTH, DERBYSHIRE.
The Great Stove was erected in 1836–40 to the design of Joseph Paxton, head gardener of Chatsworth, assisted by Decimus Burton. It covered three-quarter of acre. Having experimented since 1828, Paxton used a ridge-and-furrow shape for the roof. The cost of construction was £33,099.10s.11d.

Inside the conservatory, birds were in the trees and fish filled the pools below. The central aisle of the Stove was of sufficient width for Queen Victoria and her party to drive through in open carriages. The interior was lit by 12,000 lamps for the occasion. The Duke of Wellington noted: 'I have travelled Europe through and through, and witnessed many scenes of surpassing grandeur on many occasions, but never did I see so magnficent a coup d'oeil as that extended before me.'

It required five explosions to demolish the Great Stove in 1920.

15 (*overleaf*) Photographer unknown, *c*. 1880s. Copyright Richard Booth.

16 (*overleaf, inset*) From a Stereoscopic Card, *c*. 1860. Copyright Höwarth-Loomes Collection.

17 (*opposite, above*) CONSERVATORY, WELBECK ABBEY, WELBECK, NOTTINGHAMSHIRE.
The conservatory was part of a massive underground building scheme initiated by the 5th Duke of Portland. The Duke was obsessional about not being seen. He travelled in a wagonette with the blinds drawn. At his London home, Harcourt House, he had erected an iron and glass screen to avoid his garden being overlooked.

At Welbeck the Duke's desire for privacy resulted in the construction, commencing in 1860, of fifteen miles of underground tunnels and connecting rooms which included a library, a billiard room, a huge ballroom, the famous riding school and gallop and the glass-roofed conservatory of *c.* 1875. Herman Muthesius mentions in *Das Englische Haus* (1904–5) that Welbeck had one greenhouse that housed nothing but a particular strain of carnation. Photographer unknown, *c.* 1900. Copyright Michael Brand.

18 (*opposite, below*) THE CONSERVATORY, DOGMERSFIELD PARK, DOGMERSFIELD, HAMPSHIRE.
A number of follies was erected in the gardens at Dogmersfield by the owner, Paulet St John, in the mid 18th century. Unaccountably all except one, the Hunting Lodge, recently the home of John Fowler the interior decorator, were demolished by Paulet St John's son between 1790 and 1800.

The arbour on the left of the photograph may be an attempt to emulate the style of the earlier Gothick Arch that was in the grounds. The conservatory, which is placed at some distance from the house, in addition to its more practical function, closes the view. Photographer unknown, *c.* 1900.

19 (*above*) THE CONSERVATORY, CULFORD HALL, CULFORD, SUFFOLK.
The first Culford Hall was built in 1591. It was completely rebuilt in 1796 for the 1st Marquis Cornwallis by Robert de Carle the younger from Bury St Edmunds. The house was altered in 1819 by Lewis William Wyatt and enlarged in 1889 shortly after purchase by the 5th Earl Cadogan.

Probably the conservatory was added by Earl Cadogan. The Cadogan family sold the house in 1935 to the Methodist Education Committee for use as a public school. Photographer unknown, *c.* 1910.

20 (*overleaf*) GLASSHOUSE, INWOOD, HENSTRIDGE, SOMERSET.
The estate was owned by Mr Methyr and Lady Theodora Guest. The family ironworks had started *c.* 1758 at Dowlais, Glamorgan, and continued until 1930. By then the name of the firm was Guest, Keen and Nettlefold. Mr Guest's christian name reflects the family connection with the northern part of Glamorgan. Lady Theodora, his wife, was the youngest daughter of the Marquis of Westminster.

The photograph shows the cruciform barrel-vaulted glasshouse to which was attached a more conventional greenhouse. Its great height was presumably to accommodate tall palms. The lead bird in the foreground exemplifies Mr Guest's fondness for lead figures. Day, *c.* 1910.

21 THE CONSERVATORY, CAPESTHORNE HALL, SIDDINGTON, CHESHIRE.
The water-colour, at Capesthorne Hall, of the conservatory was made by James Johnson c. 1845.
A booklet entitled *Whitsuntide Ramble to Capesthorne Hall*, published c. 1848 for the benefit of the
Macclesfield Public Baths and Washhouses, described the building:

'This beautiful Conservatory was erected under the superintendence of Mr Paxton. It does infinite credit to the taste and judgement of himself and his late employer... the Conservatory is separated into three divisions or avenues, each affording a delightful walk for him who is to meditative mood inclined. The centre division is the loftiest, being in height 25 feet; the length of the Conservatory is 150 feet, and the breadth 40 feet; at the end of the centre division is a Stove Conservatory, entirely devoted to tropical fruits and plants.

In the centre division of the Conservatory, and on your left hand is a splendid specimen of the Aubutilon Striatum, the finest in the Kingdom; on your right is a large plant of the Datura Lutea, or yellow trumpet flower; the roof is adorned with beautiful and magnificent creepers; those most worthy of your notice are Passifloras, specimens of the Beaumontia Grandiflora, Cobea Scandens, Hardenbergia, Grandiflora Roses, Budlea, Salvifolia etc, whilst throughout the whole there are unequalled specimens of the Acacia.

In the centre bed of the Stove Conservatory, you will observe specimens of the Bread Fruit, Coffee, Sugar, Musa and other rare plants.'

William Bromley Davenport, MP for Macclesfield and a keen actor, organized family theatricals in a theatre that he had converted from part of the stable wing. It opened in January 1890, and afterwards the audience was served refreshments in the conservatory. The present Lady Bromley Davenport describes the scene in her *History of Capesthorne* (1974): 'and there, in a magical world of Chinese lanterns, exotic flowers, and intoxicating scents, they sipped their wine and laughed and talked, protected by the fragile glass from the cruelty of the winter night'.

The Conservatory was demolished in 1920. Herbert Felton, 1954.

WINTER GARDEN, EASTWELL PARK, EASTWELL, KENT.

22 Eastwell Park was inherited by the 3rd Lord Gerard in 1902 at the early age of nineteen. The house was demolished in 1926 and a new house was later built on the site. Photographer unknown, c. 1904. Copyright Captain G. Brodrick.

23 The photograph shows a house party on the steps of the winter garden. Edward VII and Mrs Keppel are among the guests. Photographer unknown, 2nd July 1904. Copyright Captain G. Brodrick.

His Majesty King Edward VII at Eastwell House
July 2nd 1904.

LADY NORREYS — Hon. Mrs GEORGE KEPPEL — Hon. H. MILNER — Hon. H. LEGGE — Sir ERNEST CASSELL
Mr. L. de ROTHSCHILD — EARL of MAR & KELLIE
Mrs. L. de ROTHSCHILD — MARQUIS de SOVERIL — COUNTESS MAR & KELLIE
LORD GERARD — Hon. Mrs LOWTHER
BARONESS de FOREST — HIS MAJESTY KING EDWARD VII — LADY GERARD — LORD CHAS. MONTAGUE
BARON de FOREST — Master A. LOWTHER — Miss GOSSELIN — COUNT MENSDORFF — LADY MAUD WARRENDER
MASTER EDMONSTONE

24 (*opposite, above*) WINTER GARDEN, HALTON, BUCKINGHAMSHIRE.
Halton was a sumptuous house, in the style of a French château, designed by W. R. Rogers in 1882–8 for Baron Alfred Charles de Rothschild. His father, Lionel, had bought the estate in 1853 from the Dashwood family. Alfred, Director of the Bank of England 1868–90, was an extravagant and generous host. Mrs Clement Scott wrote in *Old Days in Bohemian London* (1919): 'no visitor ever went empty away from Mr Alfred's house in the country. Great boxes of hot-house flowers and great baskets of luxurious fruits, delicious cakes and chocolates were packed in every carriage with the departing friend. Everyone was treated alike, from a Grand Duchess to a simple Madame'.

Next to the domed winter garden was a skating rink. After acquiring the house, the RAF built a new west wing in 1935–7 which, sadly, replaced the winter garden. Photographer unknown, *c.* 1900.

25 (*opposite, below*) WINTER GARDEN, HURSLEY PARK, HURSLEY, HAMPSHIRE.
Additions were made to the early 18th century house when Sir George Cooper, created baronet in 1905, bought Hursley in 1902 for £230,000. The architect of the alterations was A. Marshall-Mackenzie, a relative.

This domed winter garden lacks the exuberance of those of the mid-Victorian period. The severe neo-Classical style is relieved by a minimum of plants, all of which were green. A solitary chair, a bronze dog and the central fountain and pool are the only additions. Bedford Lemere, 1905.

26 (*above*) THE HALL, OLD CATTON, NORFOLK.
The Hall became the home of Samuel Gurney Buxton when he bought the late Georgian house in 1876. He was JP, Deputy Lieutenant, and High Sheriff of Norfolk in 1891.

The winter garden is a somewhat exotic creation for its position on the outskirts of Norwich. E. C. Le Grice, *c.* 1920.

27, 28 (*opposite and above*) WINTER GARDEN, SOMERLEYTON HALL, SUFFOLK.
Somerleyton was remodelled in 1844–51 by its owner, Sir Morton Peto, an entrepreneur who had built railways all over the world. He was a contractor for the Houses of Parliament where he encountered John Thomas, a sculptor and mason, whom he appointed to be his architect for Somerleyton.

Financial difficulties forced Sir Morton to put the house on the market in 1861, and in 1863 it was bought by a newly created baronet, Sir Frank Crossley. The Crossley family had made a fortune from carpet manufacture in Halifax, where they were known for their generosity in public works. Sir Frank had previously lived in a house in Halifax, Belle Vue, which was remodelled by G. H. Stokes, and had a charming if modest conservatory. Belle Vue was sold upon his death at the age of fifty-five by Sir Frank's widow, who preferred life on a country estate.

Opinions differ as to the exact size of the Somerleyton winter garden but unquestionably it was large. Robert Kerr writing in *The Gentleman's House* (1871) described it as 'the redeeming feature' of Somerleyton which 'compensates for much that is amiss'. A fountain stood in the centre surrounded by four large beds of plants around which were walkways. The entrance was along a passage from which opened the doors to the billiard room and drawing-room. This arrangement avoided the danger of damp air reaching rooms in the house. Thomas brought sculpture and vases from Italy and Germany to decorate his creation.

The winter garden was demolished in 1912. Newton, *c*. 1900.

29 (*overleaf, inset*) 165 DENMARK HILL, DULWICH, LONDON.
Sir Henry Bessemer, the inventor and engineer, moved to this address next door to John Ruskin in 1863 and began an enormous programme of building and refurbishment on that which might be called, despite its comparatively small size, his estate. Deer roamed the lawn.

The cast-iron winter garden was designed by Bessemer, with the assistance of Banks and Barry, and made by Andrew Handyside and Co. It brought a glimpse of the Alhambra to Denmark Hill. Blanch in his *The Parish of Camerwell* (1875), mentions the glittering dome of the conservatory which 'suggests ideas of eastern romance'.

30 (*overleaf*) The interior did not disappoint. The decoration was a profusion of gilt, soft-grey tinted glass, coloured marbles and Arabian ornament providing an exotic backcloth for the display of Minton majolica and Sèvres vases, majolica flower baskets, bronze 'Roman' gas lamps and candelabra. The floor was encaustic tiles and tesserae of quiet colours 'so as not to interfere with the brilliant colours of the flowers'. Mosaics representing the Four Seasons were set into the floor and a fifth representing Old Time was situated near the entrance. Four Sèvres china boys, representing Love, Pleasure, Folly and Repose, were sited in niches at each corner of the central area. The plants must almost have been irrelevant.

The house was demolished in 1949. Photographer unknown, *c*. 1880

31 WINTER GARDEN, SAVOY HOTEL, STRAND, LONDON.

The winter garden was an integral part of most grand hotels in the later Victorian period. They were decorated with large palms and used throughout the day as a room in which to take drinks and coffee, for reading, or listening to the characteristic small orchestras.

The winter garden at the Savoy was not installed until 1911, a surprisingly late date, and the decoration perhaps owes a little to the influence of the New World, of New York or Chicago. The *treillage* is interlaced with a profusion of fake roses rising from pots disguised by (probably gold) painted wicker covers. The chandeliers are wound with flowers in a style similar to that in the winter garden of the Ritz. The trellis pattern of the walls and ceiling is echoed in the pattern of the carpet. Doors lead to a balcony decorated with palms. Bedford Lemere, 1911.

32 STONE-FRUIT HOUSE, HIGH LEGH, CHESHIRE.
The nectarines are growing in one of a series of stone-fruit houses. The trees could be trained on a trellis, as in the photograph, or grown in bush form.

The ideal placement for a lean-to greenhouse was and is, against a south-facing wall so that the sun warmed the east end of the house in the morning, to the whole house at midday, and then warmed the west end of the building in the afternoon. Photographer unknown, *c.* 1900.

Copyright C. L. S. Cornwall-Legh.

33 GATTON PARK, REIGATE, SURREY.
Gatton, formerly the home of the Monson family, was bought in 1888 by Jeremiah Colman, created baronet in 1907. From 1899 to 1946 Jeremiah Colman was chairman of J. and J. Colman Ltd of mustard fame. He and his ten brothers played cricket as 'the XI Colman brothers'.

While Lady Colman preferred carnations, Sir Jeremiah was famous for his orchid collection and his hybridizing experiments produced a number of varieties.

Of the tropicals in the photograph, the most interesting belong to the genus *Nepenthes*, the pitcher plants. The long tubular head is lined with a sticky substance which catches insects. Pitcher plants were often grown in association with orchids for they both require a heat of about 80 – 90 degrees. As Shirley Hibberd says in *The Amateur's Greenhouse and Conservatory* (1897): 'some think no place hot enough for pitcher plants'. He recommends keeping them cool in winter to rest as they do in their native bogs.

Gatton Park was gutted by fire on 4 February, 1934, but rebuilt to the design of Sir Edwin Cooper in 1936. Newton, *c.* 1900.

34 (*opposite*) A VINERY, URPETH LODGE, URPETH, CHESTER-LE-STREET, COUNTY DURHAM.
Urpeth Lodge was owned in 1890 by Major William James Joicey, Deputy Lieutenant and MP. The Joicey family had made money from coal mines in the area in the 19th century.

The photograph shows two connecting vineries. Vines were first grown under glass in the early 18th century. It was the custom to bring vines on at different times so that grapes were available over a longer period. Walter Nicol, writing in 1802, supported the forcing of fruit: 'May not the wealthy eat a melon, peach, bunch of grapes, a pine-apple, with as much propriety as drink a bottle of port, claret, champaign [sic] or madeira?. . . Have they not the most rational amusement in the production of them?'

Hot-water pipes and ventilating windows can be seen at the right of the photograph. Beneath the vines are palms, ferns, and azalea and fruit trees in large terracotta pots. Bedford Lemere, 1890.

35 GREENHOUSES, HEWELL GRANGE, TARDEBIGGE, HEREFORD AND WORCESTER.
The present Grange was built by the 14th Lord Windsor, 1st Earl of Plymouth of the second creation in 1905. The old Hewell Grange had been dismantled and a new house built in 1884–91 to the design of Thomas Garner.

A series of connecting greenhouses are shown in the photograph. Melons are growing up the roof (they could also be reared in pits). Loudon had stated in *Remarks on the Construction of Hot Houses* (1817) that fruits for ripening must be kept very near the glass.

A few years earlier Walter Nicol, writing in *The Forcing Fruit and Kitchen Gardener* (1802) on melon growing in England, said 'the cultivation of this much-esteemed fruit is so general, that hardly a garden is to be met with where it is not followed in a greater or less degree, and that too with general success'. Gilbert White of Selborne was a keen melon grower.

The bed below the melons in the photograph is in use for growing palms and foliage with variegated leaves. A vinery is glimpsed beyond. Bedford Lemere, 1892.

36 GARDENERS, CASTLE ASHBY, NORTHAMPTONSHIRE.
A seat of the Marquis of Northampton.

Although known as the Orangery, probably because its external appearance resembled an 18th century orangery, the building shown here housed numerous types of exotic plant including a date palm, giant ferns, lilies, plumbago and pelargonium.

The gardeners photographed by the pool are moving potted palms. A gardener of a house possessing a conservatory would have spent a large proportion of his working hours taking potted plants back and forth from greenhouse to conservatory. Only the large background foliage creepers on pillars were permanent fixtures.

An estate of any size would include an office for the head gardener and areas for repotting, stores for seeds, fruit and root vegetables, hot-bed frames and space for fuel and compost heaps in addition to several, usually connecting, greenhouses.

At Bicton, Devon, gardeners' working hours were 6.00 a.m. to 5.00 p.m. Fines were introduced to discipline bad behaviour:

Coming to work on Monday morning with a dirty shirt	3d
Smoking a pipe of tobacco in the hours of work	4d
Neglecting to do a job after having been told of it the second time	3d
Any man found at his work intoxicated shall forfeit his days wages and be otherwise dealt with as thereafter shall be considered	6d
Swearing or making use of bad language, for every separate evil expression	3d
Damaging or in any way mutilating or defacing the above rules	12d

Photographer unknown, 1889.

37 (*overleaf*) FERNERY, THORNTON MANOR, BEBINGTON, MERSEYSIDE.
The manor was the home of W. H. Lever, the 1st Lord Leverhulme. He was tenant of the house from 1888 and bought it from Sir Thomas Forwood in 1891.

Mr Lever, who made soap in Warrington, created Port Sunlight (named after the brand name of the soap) in 1888 on marshes near Bromborough Pool. According to an article in *The Times*, written in 1938 on the golden jubilee of the factory and village, 'Port Sunlight . . . . was the first example of a complete plan for an industrial garden village.'

Lever also altered the village of Thornton Haugh, which was close to Thornton Manor. He built a fernery at the Manor as he had done at several of his other houses. He said: 'I had no idea that the love of ferns entitled a man to the jaw breaking title "Pteropodist" [sic] I have always been a lover of ferns . . . I feel now that I must have been a Pteropodist all my life without knowing it, like the man who found he had been speaking prose all his life'. Bedford Lemere, 1903.

38 HOTHOUSES, HOLLY LODGE, ECCLESALL, SOUTH YORKSHIRE.
Holly Lodge was a comparatively small unpretentious villa designed by Charles Hadfield, a Sheffield architect (the firm still exists as Hadfield, Cawkwell, Davidson and Partners). The owner of the house was F. A. Kelley, managing director of a firm of brewers, Whitmarsh, Watson and Co., Ltd. Until 1896 he lived at 29, Collegiate Crescent, Sheffield, where he had presumably prospered sufficiently to move to a then country estate.

It is difficult to imagine how a comparatively small house could absorb this huge collection of background foliage. Plants with variegated and ornamental foliage, such as the tradescantias and members of the arrowroot family from tropical America, were immensely popular with the Victorians. Bedford Lemere, 1897.

39 (*opposite, above*) THE ORANGERY, HOLLAND HOUSE, LONDON.
Originally the home of Sir Walter Cope, Chamberlain of the Exchequer and Keeper of Hyde Park, Holland House was earlier known as Cope House or Cope Castle. In 1629 the house passed to Sir Henry Rich, 1st Earl of Holland, who had married Walter Cope's daughter.

The room now known as the orangery was not converted to that use until the 19th century. It was originally just a part of the 1638–40 stables and coach house.

When the photograph was taken the room seems to have been used as a floral route to a garden room at the end. This room, like one at Abbey House, Evesham, has a false mirror, which here allows a view through to an urn in the garden. Newton, *c.* 1910.

40 (*opposite, below*) CONSERVATORY, EATON HALL, EATON, CHESHIRE.
Eaton Hall was transformed for the 1st Duke of Westminster between 1870–83 by Alfred Waterhouse.

The conservatory had one solid wall and one glazed. The shape of the room is reminiscent of the Elizabethan long gallery and seems designed for promenades.

Most of Eaton Hall was demolished in 1961. A new Eaton Hall replaced it. Newton, *c.* 1910.

41 CONSERVATORY, SURREY HOUSE, 7 MARBLE ARCH, LONDON.
Surrey House was the town house of the first and last Lord Battersea, whose family name was Flower. Lady Battersea was the elder daughter of Sir Anthony de Rothschild, owner of Aston Clinton. The couple also owned the Plaisaunce, Overstrand, Norfolk, where they were in the heart of the artistic 'Queen Anne Revival' set.

The Batterseas lived at 7 Marble Arch which had a bedroom designed by Bugatti, for thirty years. The house was a meeting place for members of the Liberal party, musicians, artists, writers and philanthropists. Lady Battersea, who did enormous numbers of good works, was disappointed not to settle in Battersea 'among the working classes'. In her book *Reminiscences* (1922) Constance Battersea describes a social evening at which Mr Gladstone and Robert Browning were present, the latter reading his poetry aloud. On another occasion, following entertainment by a Hungarian band, Lady Hallé snatched a violin from a player and struck up the Radetsky March.

Lady Battersea mentions looking from a window of Surrey House in 1907 and seeing the German Emperor and Empress driving in semi-state to the Guildhall to receive the freedom of the City. She particularly noted the Empress's 'splendid pearl necklace'.

The conservatory is clearly designed for display. Without the enormous quantities of chrysanthemums in pots the room would have been almost empty. James W. L. Hyatt, *c.* 1890. Copyright Christopher Wood Gallery.

42, 43 (*opposite, above and below*) **WINTER GARDEN, CHATEAU IMPNEY, DROITWICH, HEREFORD AND WORCESTER.**
The château, in the style of Francis I and Louis XIII of France, was designed by a Frenchman, Auguste Tronquois, in 1875 for John Corbett MP, who made a fortune from Droitwich salt. It is alleged that Corbett wanted to remind his French bride of her homeland. The cost of construction was £247,000.

The winter garden, which was made by Boulton and Paul, was not attached to the house.

Externally it was like an enormous greenhouse; but inside it was almost a forerunner of Disneyland. Bedford Lemere, 1892.

44 (*previous spread*) CONSERVATORY, HOLLY LODGE, ECCLESALL, SOUTH YORKSHIRE.
A pile of rocks was used decoratively and to conceal the flower pots. The display seems not to have featured alpines, as one might expect, but orchids. Bedford Lemere, 1897.

45 CONSERVATORY, WYNYARD PARK, GRINDON, CLEVELAND.
The Conservatory was designed by Benjamin Wyatt in 1822 for the 3rd Marquess of Londonderry. It was altered inside later but remained a magical place in which to walk or sit, as the *chaise longue* to the left of the photograph attests. Electric light made evening promenades possible. As was usual in conservatories, the decorative iron grilles concealed the heating pipes and allowed excess moisture to escape.

All the flowers and most of the foliage plants are in pots. In the foreground are lilies, carnations and *companula pyramidalis*. Newton, *c.* 1900.

46 WOOD, END, SCARBOROUGH, NORTH YORKSHIRE. Wood End was the home of Sir George and Lady Sitwell, the latter continuing to live there after Sir George's death. He had been a keen gardener and was the author of *On the Making of Gardens*. However, according to Sir Osbert, one of his sons, 'no man knew or cared less' for flowers. Lady Sitwell liked scent and colour in a garden but hated 'Horticulturalist's blossoms'. The unusual double-storied conservatory was apparently a felicitous way of linking two buildings.

47 The charming water-colour of the conservatory was painted by F. Breary Robinson in 1893. The trellis on the end wall no longer survives but the conservatory, still filled with plants, is appropriately part of the Natural History Museum. NMR, 1981.

48, 49 CONSERVATORY, CARROW HOUSE, NORWICH, NORFOLK.
Carrow House was the home of Jeremiah James Colman, MP for Norwich. He began to move the family mustard business from Stoke Holy Cross to Carrow in 1854.

Jeremiah James Colman, a man of liberal, non-conformist disposition, initiated many schemes for the education and welfare of his workers and their dependants. He donated to the city of Norwich books and other material relating to the history of Norfolk and, in addition, most of his collection of paintings by the Norwich School.

Jeremiah James was a naturalist and loved flowers. In addition to greenhouses in the grounds, he erected a conservatory at Carrow House made by Boulton and Paul. The conservatory is made of teak and brick and there are some charming 'Arts and Crafts' style stained-glass windows. NMR, 1981.

50 64 OLD CHURCH STREET, CHELSEA, LONDON.
The house was designed in 1936 by Erich Mendelsohn and Serge Chermayeff. Next door was a house by Walter Gropius. Together they formed a 'spearhead group' of 'Modern Movement' houses in the heart of traditional Chelsea.

A squash court is included in the house plan of No 64.

The 'clean lined' conservatory was added in 1963. NMR, 1981.

51 (*opposite*) CONSERVATORY, BISHOPSTONE HOUSE, BISHOPSTONE, WILTSHIRE.
The house was designed by J. Lowder of Bath in 1820 for the Reverend Thomas Bromley. Completion took place while a Reverend Montgomery was the incumbent. Bishopstone House is said to be the original of Trollope's Plumpstead Episcopi.

The conservatory, which leads from the drawing-room, was added in 1828. It would seem that the gentry were not slow to emulate architectural fashions adopted by the aristocracy. NMR, 1981.

52, 53 (*overleaf*) THE WINTER GARDEN, SHRUBLAND HALL, BARHAM, SUFFOLK.
The winter garden was added to Shrubland Hall in 1830 by Gandy-Deering, an associate in Greece of Lord Elgin. He was a pupil of James Wyatt.

   In 1860 Shrubland was inherited by Rear-Admiral Sir George Broke-Middleton. He was the fourth son of the famous Admiral Sir Philip Vere Brook who, in 1813, commanded the *Shannon* and captured the American frigate *Chesapeake* in Boston Harbour. The figurehead of HMS *Shannon* is by the steps leading from the winter garden to a small drawing-room.

   Shrubland Hall is the home of Lord and Lady de Saumarez, who run the house as a health farm. Sir Philip Broke's granddaughter was a Lady de Saumarez. The de Saumarez family originated from Guernsey and has a nautical tradition. NMR, 1981.

54 HORNCLIFFE HOUSE, HORNCLIFFE, NORTHUMBERLAND.
The house was built of pink sandstone c. 1810.

The winter garden leads from the morning room. 'Barley sugar' columns support the roof structure.

On one wall is a niche flanked by 'Florentine' columns and topped by incised decoration in similar style.

The plants include a vine, plumbago, mimosa, camellias, fuchsias, tree peonies and arum lilies. NMR, 1981.

55 CONSERVATORY, FAIRLAWNE, PLAXTON, KENT.
Formerly, Fairlawne was the home of Mr Peter Cazalet, who trained National Hunt horses for HM The Queen Mother.

The conservatory was equipped with delightful wicker chairs from the Far East. There was a display of sculpture among the plants.

The room was converted to an open loggia in the 1950s. Newton, c. 1910.

56, 57 CONSERVATORY, MOUNT TAVY, TAVISTOCK, DEVON.
This late Georgian house with Victorian additions on the east bank of the River Tavy was bought by a Devon businessman, Daniel Radford. He had previously built, in 1872, Lydford Bridge House on the Lydford River, about 6 miles north of Tavistock.

The conservatory is thought to have been built *c.* 1875. At one end was a circular area with seats for conversation.

The house became a boy's preparatory school in the inter-war period. Photographer unknown, *c.* 1900. Copyright C. A. Ralegh Radford.

58 CONSERVATORY, SOUTHOVER HALL, BURWASH, EAST SUSSEX.
The ownership of Southover Hall by Mariano de Murietta is first noted in *Kelly's Directory* for 1878.

Mr Murietta was a banker and friend of Joseph Duveen, the art dealer. He suggested to Duveen that old Spanish families living in Mexico and impoverished by the revolution against Spain were likely to wish to sell fine works of art. Duveen sent his son Charles to Mexico to investigate.

The conservatory was quite clearly for sitting in. The tableful of pelargoniums (which originate from South Africa) in the foreground relieves the overall impression of green foliage. There is no sign of the popular plant from Murietta's homeland *Choisya ternata* (the Mexican orange flower). Bedford Lemere, 1893.

59, 60 (*opposite*) CONSERVATORY, CHATEAU IMPNEY, DROITWICH, HEREFORDSHIRE AND WORCESTERSHIRE.
The conservatory at Chateau Impney was added to the house *c.* 1891. It was designed by J. R. Nicholl.

The tiled walls in 'French' style perhaps reflect the influence of Mrs Corbett, the owner's wife, who was a Frenchwoman. The generally green display, like that at Somerleyton Hall, Suffolk, provides a foil for an exhibition of sculpture.

In 1962 it was reported that Billy (later Sir William) Butlin was interested in providing a 'playground for the Midlands' at Chateau Impney, which had been owned for eighteen years by a Mr Ralph Edwards. It was then, and still is, an hotel. Bedford Lemere, 1892.

61 (*above*) CONSERVATORY, THE HOLMSTEAD, MOSSLEY HILL ROAD, LIVERPOOL, MERSEYSIDE.
The house, which was the home in the 1880s of the Holt family, who owned a shipping line, was later given by them to Liverpool Corporation.

This conservatory, though perhaps a particularly pretty example, exemplifies those added to the majority of middle-class houses between 1880 and the First World War. The Gothic design of the conservatory blends well with the house to which it is attached. There would be room for a changing display of plants and a few chairs. NMR, 1967.

62 (*overleaf*) CONSERVATORY, WINTERFOLD, CRANLEIGH, SURREY.
Winterfold, designed by E. l'Anson, was the home of a widower, Sir Richard Webster, created Lord Alverstone in 1900. He was a barrister, MP for Launceston and the Isle of Wight, Attorney-General, Master of the Rolls and Lord Chief Justice.

The conservatory led from the drawing-room. It was fitted in the manner of a greenhouse with shelves for the display of plants round the room. Wooden fretwork panels concealed the heating pipes. Cineraria, daisies and arum lilies are among the plants. Bedford Lemere, 1890.

63 (*opposite*) CONSERVATORY, HIGH LEGH, CHESHIRE.
This photograph of the Conservatory erected in the early 1890s indicates that a number of conservatories were used as floral sitting-rooms. The seated couple are Colonel H. M. Cornwall-Legh and his wife. This room, like many conservatories dating from the turn of the century, is not of sufficient size to permit promenades.

   High Legh was demolished in 1962. Photographer unknown, *c.* 1910. Copyright C. L. S. Cornwall-Legh.

64 CONSERVATORY, FULWELL PARK, RICHMOND UPON THAMES, GREATER LONDON.
Fulwell Park was occupied by King Manoel of Portugal after his exile from Lisbon in 1910.

   The charming conservatory, though rather short of plants, has great atmosphere. It provides a setting in which, one feels, characters from a story by Somerset Maugham would feel at ease.

   The house has been demolished. Newton, *c.* 1910.

66 (*below*) CONSERVATORY, THE ELMS, SPANIARD'S ROAD, HAMPSTEAD, LONDON.
The Elms was the home of Joel Joseph, later Sir Joseph, Duveen, the famous decorator and dealer in *objets d'art*. Sir Osbert Sitwell described Duveen as having 'a clownish amiability, which resembled that of a Jewish tumbler on the music hall stage'. After a shaky start, Duveen became 'the greatest salesman of his time' with a shop in London and branches in Paris and New York.

The conservatory was of marble and wood. Bedford Lemere, 1899.

65 (*above*) CONSERVATORY, BELLMOOR, EAST HEATH ROAD, HAMPSTEAD, LONDON.
The house was designed by C. E. Birch for Thomas J. Barratt, the antiquary and joint head of Pear's Soap. Barratt wrote a three-volume local history entitled *Annals of Hampstead* (1912) but his main claim to fame was that he was the man who devised the 'Bubbles' campaign for Pears. He was often referred to as the father of modern advertising. Having revolutionized Pears when he became a partner after his marriage to Francis Pear's daughter, he persuaded Lily Langtry and other household names to endorse Pears Soap; they were not paid a fee. In 1891 he introduced the *Pears Annual*.

Barratt's conservatory was in an eclectic taste that reflected his diverse interests.

Bellmoor has since been demolished. Bedford Lemere, 1893.

67 (*opposite*) CONSERVATORY, 45 GROSVENOR SQUARE, LONDON.
A 'classical' conservatory designed by Mellier and Co., for Sir James Miller, the racehorse owner. The house, built in 1725, was first occupied by the Marquess of Blandford, MP. Edward Wimperis redesigned the façade in 1902; the house has since been demolished.

In the late-Victorian period Grosvenor Square was a favoured location for the homes of the newly-rich and successful. A building contractor, a shipping magnate, a Canadian company director and several financiers were among the residents. Bedford Lemere, 1897.

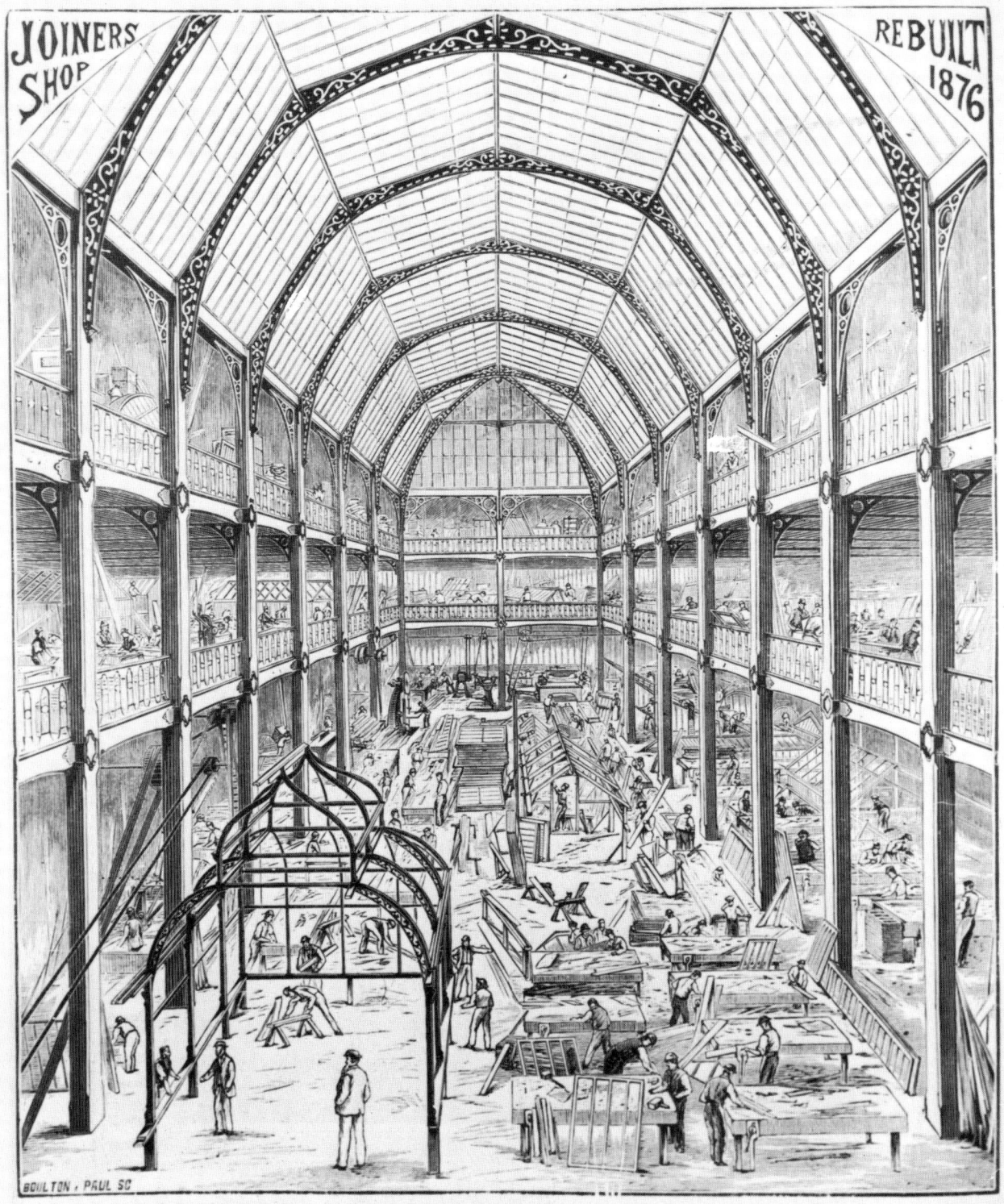

68–77 BOULTON AND PAUL LIMITED, NORWICH, NORFOLK.

An enormously increased demand arose for conservatories in the late Victorian period. Boulton and Paul of Norwich, which had originated as an ironmongers business in 1797, was one of several firms that gave particular emphasis to the manufacture of glasshouses. Boulton and Paul's first catalogue was issued in 1875. The huge diversity of goods that were produced from the Rose Lane Works included, in addition to conservatories and glasshouses, garden houses, and bungalows for use as officers' quarters in the Boer War.

During the First World War, the firm expanded to produce aircraft, necessitating the removal to a new site in 1916. The Rose Lane Works were sold by the Company in 1922, and the site is at present occupied by a discotheque. NMR, 1981. Copyright Boulton and Paul Limited.

ROSE LANE WORKS, NORWICH.   33

PLANT HOUSES, VINERIES, &c.

ORCHID HOUSES

Of every description erected and fitted complete with Heating Apparatus, Tanks, Stages, Trays, and Shelvings.

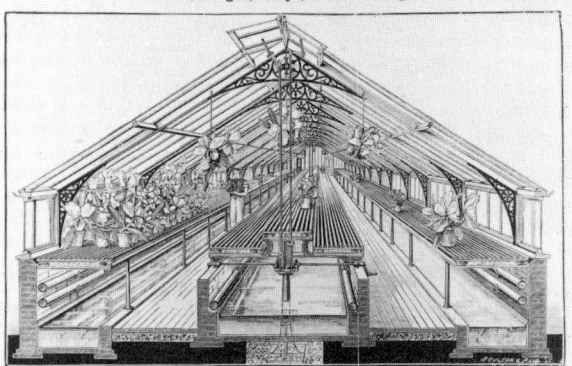

Interior View of Span-roof Orchid House at Downside, Leatherhead.

The celebrated Orchid Houses of

W. LEE, ESQ., DOWNSIDE, LEATHERHEAD, SURREY.
C. W. WALKER, ESQ., MILNTHORPE, WESTMORELAND.
W. SPINDLER, ESQ., OLD PARK, ISLE OF WIGHT.
W. VANNER, ESQ., CHISLEHURST, KENT.
SIR W. HUTT, APLEY TOWERS, RYDE, ISLE OF WIGHT,

and many others in the kingdom have been erected by us.

GENTLEMEN,— PARKHEAD, MILNTHORPE.
The Orchid Houses built by your firm at Brettargh Holt, for C. W. Walker, Esq., have been fully tested by me, and I am pleased to say the whole gives entire satisfaction. The Heating Apparatus is most complete; I have worked both boilers, together and separately, and they do the work in a most satisfactory manner with very little trouble.
(Signed) ALEX. MAC GREGOR.

ROSE LANE WORKS, NORWICH.   31

SMALL GREENHOUSES

SUITABLE FOR

VILLA RESIDENCES, COUNTRY HOUSES, &c.

THIS series of Greenhouses, Nos. 44 to 49, embrace those useful for small requirements. In all our quotations we give the price of the finished article, fitted, painted, and glazed all ready for putting together, and not, as usually advertised, with glass cut up and packed, etc., leaving half or more of the work to be done on delivery. From the many testimonials we have received and give for reference, it will be seen that every satisfaction is given, that trouble is avoided, and the houses themselves give pleasure at once, as in a few hours they are quite ready to receive the plants. Upwards of 800 of these houses have been supplied in a few years, mostly from recommendation, as the houses are never specially advertised.

Our Heating Apparatus for these Houses are simple in make, easy to work, and will last a lifetime without repairs, see page 70. Our object is to supply only those articles which from experience and constant use we have proved will give satisfaction.

Stages and paths are quoted for, and are all priced fearless of competition.

View of Span-roof Greenhouse No. 52, with end removed.

Showing a very useful arrangement of our No. 62 Lean-to Pits on either side of the house. Can be applied to Nos. 50, 51, or 52 Greenhouses. A great deal of room can be gained if the base of the Greenhouse is omitted; it is then arranged to stand on piers of brickwork, or, if Tenant's Fixture, on iron stands, at the same price.

BOULTON AND PAUL, MANUFACTURERS,   44

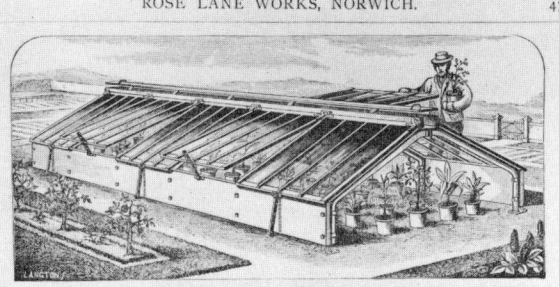

No. 62. LEAN-TO FRAME OR FORCING PIT.

SPECIFICATION.—The lights are made of thoroughly seasoned red deal, glazed with 21 oz. English sheet glass, including ends. They are hinged at the top, as shown, so that they can be set open for ventilation in a simple way; each light is also arranged so that it can be easily lifted off when required, the hinges being specially constructed to allow of this being done. The whole painted four coats of good oil colour. This form of frame can be placed in any convenient position by the side of a greenhouse or any wall.

Plans of requisite brickwork sent on receipt of order.

Cash Prices, Carriage Paid.

Length.	Width.	Height at back above wall.	Cash Prices, Carriage Paid.
			£ s. d.
10 ft.	3 ft.	1 ft. 9 in.	2 15 0
15 ft.	3 ft.	1 ft. 9 in.	3 15 0
20 ft.	3 ft.	1 ft. 9 in.	4 15 0
25 ft.	3 ft.	1 ft. 9 in.	5 15 0
30 ft.	3 ft.	1 ft. 9 in.	6 15 0
36 ft.	3 ft.	1 ft. 9 in.	7 15 0

Glazed Division, each 8/-

10 ft.	4 ft.	2 ft. 0 in.	3 7 6
15 ft.	4 ft.	2 ft. 0 in.	4 12 0
20 ft.	4 ft.	2 ft. 0 in.	5 15 0
25 ft.	4 ft.	2 ft. 0 in.	6 18 6
30 ft.	4 ft.	2 ft. 0 in.	7 18 0
36 ft.	4 ft.	2 ft. 0 in.	8 17 6

Glazed Division, each 10/-

10 ft.	5 ft.	2 ft. 3 in.	4 5 0
15 ft.	5 ft.	2 ft. 3 in.	5 12 0
20 ft.	5 ft.	2 ft. 3 in.	7 1 0
25 ft.	5 ft.	2 ft. 3 in.	8 11 0
30 ft.	5 ft.	2 ft. 3 in.	10 0 0
36 ft.	5 ft.	2 ft. 3 in.	11 10 0

Glazed Division, each 12/-

Section of No. 62 Lean-to Frame or Forcing Pit, shewn built against a greenhouse.

Estimates given for any length on application.

TESTIMONIALS.

PORTFIELD HOUSE, CHICHESTER, Dec. 8th, 1887.
GENTLEMEN,—I have received the four frames, and had them unpacked, and am greatly pleased with them. They are the best made frames I ever had, and am glad my friend, Mr. Daglish, gave me your address. Signed, JOS. DAVIES, J.P.

SULCHAY HOUSE, WISBECH, Feb. 16th, 1887.
DEAR SIRS,—I enclose P.O.O. in payment of your account for frames, which give entire satisfaction.
MESSRS. BOULTON AND PAUL.   Signed, WM. GROOM, junr.

ROSE LANE WORKS, NORWICH.   47

No. 70. UNIVERSAL SPAN ROOF PLANT FRAME.

Fig. 1. This Frame has been designed to meet the wants of those who require a useful place, capable of being turned to a variety of purposes.

The above represents the Frame set upon the general level ground, with the end removed, shewing greenhouse plants, which can be prepared for the conservatory, &c.

No. 70. ARRANGED AS A SMALL GREENHOUSE.

Fig. 2 shews the Frame with a sunk path, which may be bricked or boarded, when, with the addition of a door, it makes a useful house, and may be heated or not as required. We strongly recommend this for Auriculas. A cool bed, with plenty of drainage, is required, and no Heating Apparatus.

Examples of plant houses and frames which could be purchased from Boulton and Paul. The illustrations show the manner in which plants might be arranged.

The use of testimonials in the catalogues obviously encouraged good sales. NMR, 1981. Copyright Boulton & Paul Limited.

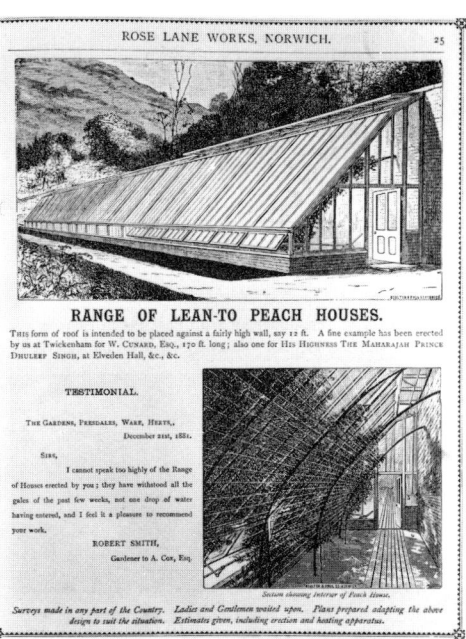

78, 79 TREE-HOUSE, THE HALL, PITCHFORD, SHROPSHIRE.
This whimsical tree-house has been dated to the 1760s but it seems ahead of its time. Inside, is a 'Gothick' plaster ceiling and cornice. A sunburst features in the centre. The half timbering on the outside walls is a 20th century addition. G. B. Mason, 1959.

80 SUMMER HOUSE, MARLBOROUGH HOUSE, FALMOUTH, CORNWALL.

The Cornish sunshine encouraged palms to flourish around this enchanting summer house. It stood in the grounds of Marlborough House, which was altered c. 1810 for Captain John Bull, who was formerly in the West India Packet Service between 1798 and 1806. Bull renamed the house after his boat *The Duke of Marlborough*, famous in encounters with privateers. A marble relief of the ship is sited on the front of the house.

The painting of the summer house was made in 1943 by Barbara Jones, author of *Follies and Grottoes* (1953). By 1949 the summer house was in ruins. Photographer unknown, 1943. Copyright Pilgrim Trust.

81 ASCOTT, WING, BUCKINGHAMSHIRE.

The house of 1606 was remodelled by the architect George Devey after Leopold de Rothschild bought it in 1874.

The room called the conservatory, probably added c. 1892, smothered in creepers with leaded light windows, seems more like a summer house. It is difficult to conceive how plants could flourish with so little light. Bedford Lemere, 1893.

82 TEA-HOUSE, EATON HALL, EATON, CHESHIRE.
The tea-house was designed in 1872 by John Douglas of Chester who was also responsible for the Aldford Lodge at Eaton Hall.
  Stacey Marks, RA, who painted a mural of the Canterbury Pilgrims in the Saloon at Eaton Hall, also decorated the interior of the Tea-house with painted tiles of birds and animals illustrating the Seven Ages of Man and humorous pictures of the Signs of the Zodiac.
  The garden fronting the Tea-house was lawn in 1901 and tulip beds in 1907. Newton, c. 1910.

83 (*overleaf*) THE PAVILION, THE HOLMEWOOD, ESHER, SURREY.
The property was the 'country' home of Alexander Alexander Ionides, the Greek Consul-General, who died in 1898.
  His town house, Holland Park, was decorated by William Morris, Philip Webb, T. Jekyll and Walter Crane. He moved in the 'Holland Park' circle which included J. S. Beale, the solicitor (owner of Standen, Sussex) and Alexander, the banker and patron of Whistler, whose family had been Rochester ship-builders (Aubrey House, Campden Hill).
  The group in the photograph is taking tea outside the 'rustic' pavilion apparently following games of cricket and tennis. Bedford Lemere, 1890.

84 SUMMER HOUSE, WARREN HOUSE, COOMBE HILL, KINGSTON UPON THAMES, GREATER LONDON.

Warren House was built in 1884 for Lord Wolverton, a member of the Glynn family who founded William and Glynns Bank. The Paget family moved to Warren House in 1907 and remained until 1954, when it was sold to Imperial Chemical Industries.

Kingston Hill had been developed by the National Freehold Land Society in 1853. Coombe was an affluent Victorian suburb. John Galsworthy was born there in 1867. The fictional 'Robin Hill', Soames' house in the Forsyte Saga, was on Coombe Hill. Galsworthy wrote in 1929 to the *Surrey Comet*: 'In those days as you probably remember Coombe was very different – very much as I describe it at the opening of the Man of Property. The site of the Forsyte House was the site of my fathers Coombe Warren and the grounds and coppice etc. were actual, but the house itself I built with my imagination.'

Sir Arthur Paget created the Japanese garden and its summer house. (The Japanese style was very popular at the turn of the century).

In common with many of the Victorian properties on Coombe Hill, the garden of Warren House has now been divided up but most of the features remain. Newton, c. 1910.

85 TEA-HOUSE, EASTON LODGE, LITTLE EASTON, ESSEX.

Easton Lodge was the home of the Countess of Warwick who later became an ardent socialist and left the house to the Trades Union Congress. The house has been demolished.

The gardens at Easton (and Warwick Castle) were designed by H. A. Peto in 1902. In addition to local workmen, about seventy Salvation Army waifs were employed to effect the transformation. The Japanese tea-house was built on piles over one of a series of old fish-ponds. Delicate hybrid water lilies were introduced at Easton and flourished despite fears that the Essex climate would prove too harsh. Newton, c. 1907.

86 SUMMER HOUSE, SHEPHERD'S HILL, HADLOW DOWN, EAST SUSSEX.

Shepherd's Hill was the home of Sir Edward Maufe, architect of Guildford Cathedral and St Columba's, Pont Street. Lady Maufe was art adviser to Heal's.

Sir Edward bought the house in 1926 and spent two years making alterations in a simple neo-Georgian style. The design of the summer house utilized old columns. NMR, 1976.

87 LOGGIA, THE HILL, NORTH END WAY, HAMPSTEAD, LONDON.
One of the homes of self-made men bordering Hampstead Heath, the Hill was the London home of Lord Lever, the soap manufacturer and MP for the Wirral (at the fourth attempt).

The loggia, extending the length of the house, allowed the enjoyment of more fresh air on the face than had hitherto been fashionable. It created less of a barrier between house and garden than would have a conservatory. Newton, c. 1910.

88 THE LOGGIA, ALDENHAM HOUSE, ALDENHAM, HERTFORDSHIRE.
Aldenham House was formerly the home of Henry Gibbs, who was governor of the Bank of England from 1875 to 1877; he was created Lord Aldenham in 1896. His third son, Vicary Gibbs, editor of *The Complete Peerage*, was a keen collector of plants and in the early 20th century he made the gardens of Aldenham famous.

The house is now Haberdashers' Aske's School.

The loggia allowed the display of classical plaques and mosaics and provided a protected place to sit while enjoying the open air. Newton, c. 1910.

89 (*opposite*) THE LOGGIA, PYRFORD COURT, WOKING, SURREY.
The loggia was built to the neo-Georgian design of Clyde Young in 1906 for Lord Iveagh, then the Hon Rupert Guinness. Lord Iveagh's wife was a member of the Onslow (Clandon Park) family.
   Lord Iveagh designed additions to the house which were begun in 1914. The outbreak of war held up the work but it was completed between 1920 and 1928.
   Pyrford Court became the headquarters of the Brigade of Guards shortly after Dunkirk.
   The gardens were the creation of Lord and Lady Iveagh. The loggia faces a lawn and was equipped with seats and a box containing croquet mallets. Newton, *c*. 1910.

90 30 OLD SNEED PARK, STOKE BISHOP, BRISTOL, AVON.
A 'Modern Movement' house designed in 1936 by Mark Hartland Thomas, a Bristol architect, for H. J. Hampden Alpass. The style, typically for houses built during the inter-war period, reflects the preoccupation at that time with the benefits to health of light and air. The *Architect and Building News* (15 October 1937) mentions how 'the french doors and verandah leading out of the hall add to the feeling of unrestricted movement'. The hall rose through two stories and was surmounted by the 'summer house' occupying part of the flat roof.
   Although the Victorians and Edwardians would have been dismayed by such departures from the contrived, protected world of the conservatory, in the 1920s a leader of 'Modern Movement', Berthold Lubetkin, cautioned, 'My personal interpretation is that these buildings cry for a world which has never come into being'. Herbert Felton, 1937. Copyright *The Architect and Building News* (B. W. S. Publishing).

91 ROOF GARDEN, DERRY AND TOMS, KENSINGTON HIGH STREET, LONDON.
This *roofless* garden room is on top of the former Derry and Toms department store and was designed by B. George in 1933. It became in turn the famous store 'Biba', and 'Régine's' and 'The Gardens' night clubs. Photographer unknown, 1930s.

INDEX

References in roman numerals refer to introduction pages. Other references refer to the plates.

Aldenham House, Herts 88
Ascott, Bucks 81

Barnes, James ii, 13
Barton Seagrave Hall, Northants 1
Bellmoor, Hampstead, London 65
Bicton House, Devon 13, 36
Bishopstone House, Wilts 51
Boulton and Paul Ltd., Norwich 68–77
Bromborough Hall, Merseyside 11
Burton, Decimus 7, 15, 16
Byram Park, N. Yorks 5

Camellia House, Wollaton Hall, Notts iii, 12
Capesthorne Hall, Cheshire iii, 21
Carrow House, Norwich 48, 49
Castle Ashby, Northants 6, 36
Chateau Impney, Droitwich 42, 43, 59, 60
Chance Bros. iii
Chatsworth, Derbys iii, 15, 16
conservatories ii, iii, 14, 17, 18, 19, 21, 40, 41, 44, 45, 46, 47, 48, 49, 50, 51, 55, 56, 57, 58, 59, 60, 61, 62, 63, 64, 65, 66, 67, 81
conservatories, decline of vi

conservatories, owners of v, vi

Denmark Hill, 165, Dulwich, London 29, 30
Derry and Toms, Kensington, London 91
Dogmersfield Park, Hants 18

East Cliff Lodge, Ramsgate, Kent 7
Easton Lodge, Essex 85
Eastwell Park, Kent 22, 23
Enville Hall, Staffs 14

Fairlawne, Kent 55
Farley Hall, Staffs 8
ferneries iv, 37
flowers v
Fulwell Park, Richmond upon Thames 64

Gatton Park, Surrey 33
Great Stove, Chatsworth iii, 15, 16
gardeners 13, 36
greenhouses 7, 8, 11, 35
Grosvenor Square, 45, London 67

Handyside and Co. 29
Halton, Bucks 24
heating ii, iii, iv
Hewell Grange, Hereford and Worcester 35
High Legh, Cheshire 32, 63
Hill, The, Hampstead, London 87
Hibberd, Shirley iv, v
Holland House, Kensington, London 39
Holly Lodge, Ecclesall, S. Yorks 38, 44
Holmstead, The, Liverpool 61
Holmewood, The, Esher 83
hot beds ii
hot walls i, ii
hothouse 38
Horncliffe House, Northumb 54
Inwood, Somerset 20

Kensington Palace, London i
Kerr, Robert ii, iii, iv, v, 27, 28

Laurence ii
loggias 87, 88, 89
Loudon, John Claudius iii, v, vi, 7, 8, 35
Luscombe, Devon ii

Mapledurham, Oxon 9
Marlborough House, Falmouth 80
Miller, Philip i
Mount Tavy, Devon 56, 57

Nash, John ii
Nicol, Walter iii, 35

Old Catton, Norfolk 26
Old Church Street, 64, Chelsea, London 50
Old Sneed Park, Bristol 90
orangeries i, 1, 2, 3, 4, 5, 6, 9, 10, 39

Palm House, Bicton House, Devon 13
Panshanger, Herts 3
Parkinson, John i
Paxton, Joseph iii, 15, 16, 21
Payne Knight, Richard ii
Pavillion, The Holmewood, Esher 83
Pitchford, The Hall, Salop 78, 79
Pyrford Court, Surrey 89

Roof Garden, Derry and Toms 91

Savoy Hotel, Westminster, London 31
Sezincote, Glos 2
Shepherd's Hill, Hadlow Down 86
Shrubland Hall, Suffolk 52, 53
Somerleyton Hall, Suffolk v, 27, 28
Southover Hall, Burwash, E. Sussex 58
stone fruit house 32
'stove' ii
summer houses 80, 84, 86
Surrey House, Westminster, London 41
Stevenson, J. ii, v

Turner, Richard 7
tea-houses 82, 85
Thornton Manor, Merseyside 37
Tree-House, Pitchford, Salop 78, 79

Urpeth Lodge, Chester-Le-Street 34

vinery 34

wardian case iv
Warren House, Coombe Hill 84
Watts ii
Welbeck Abbey, Notts 17
Webb, Philip vi
Winterfold, Cranleigh, Surrey 62
winter gardens iii, 22–31, 42, 43, 52–54
Wollaton Hall, Notts iii, 12
Wood End, Scarborough iii, 46, 47
Wrest Park Beds 4
Wynyard Park, Cleveland 45